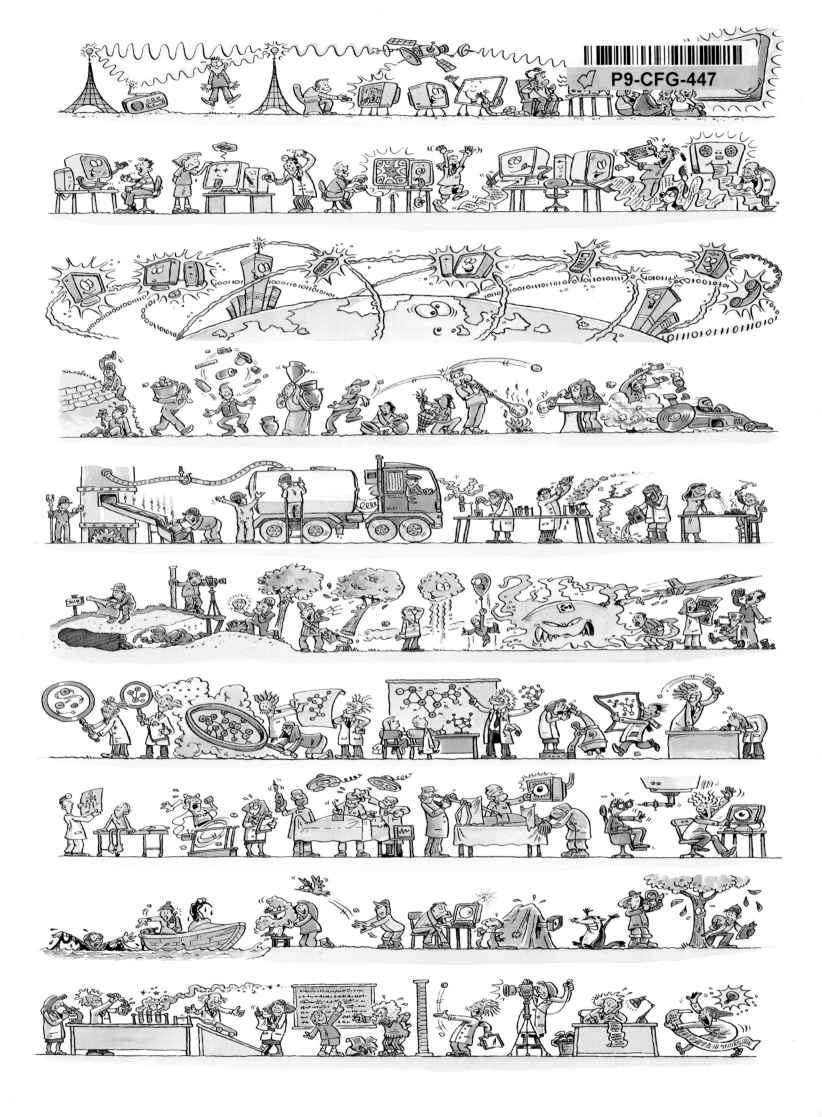

100
things you should know about
SCIENCE

100
things you should know about
SCIENCE

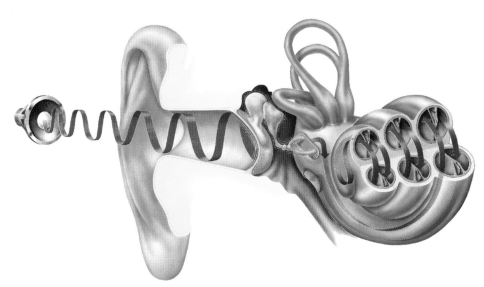

Steve Parker

Consultant: Peter Riley

Mason Crest
Publishers

Mason Crest Publishers Inc.
370 Reed Road, Broomall, PA 19008
(866) MCP-BOOK (toll free)
www.masoncrest.com
This edition first published in 2003

Miles Kelly Publishing,
Bardfield Centre, Great Bardfield, Essex, CM7 4SL, U.K.
Copyright © Miles Kelly Publishing 2001, 2003

2 4 6 8 10 9 7 5 3 1

Library of Congress Cataloging-in-Publication Data on file
at the Library of Congress

ISBN 1-59084-456-4

Editorial Director: Anne Marshall
Editors: Amanda Learmonth, Jenni Rainford
Design: Angela Ashton, Joe Jones
Indexing, Proof Reading: Lynn Bresler
Americanization: Sean Connolly

Printed in China

ACKNOWLEDGMENTS
The publishers would like to thank the following artists who have contributed to this book:

Mark Bergin   Maltings Partnership
Steve Caldwell   Helen Parsley
Richard Draper   Terry Riley
Chris Forsey   Martin Sanders
Mark Franklin   Mike Saunders
Shammi Ghale   Steve Weston
Peter Harper   Tony Wilkins
Janos Marffy

Cartoons by Mark Davis at Mackerel

www.mileskelly.net
info@mileskelly.net

Contents

Why do we need science?

1 Even one hundred books like this could not explain all the reasons why we need science. Toasters, bicycles, mobile phones, computers, cars, light bulbs—all the gadgets and machines we use every day are the results of scientific discoveries. Houses, skyscrapers, bridges, and rockets are built using science. Our knowledge of medicines, nature, light, and sound comes from science. Then there is the science of predicting the weather, investigating how stars shine, finding out why carrots are orange…

▶ In a big city, almost every vehicle, building, machine, and gadget is based on science and technology.

Machines big and small

2 **Machines are everywhere!** They help us do things, or make doing them easier. Every time you play on a seesaw, you are using a machine! A lever is a stiff bar that tilts at a point called the pivot or fulcrum. The pivot of the seesaw is in the middle. Using the seesaw as a lever, a small person can lift a big person by sitting further from the pivot.

Thread

3 **The screw is another simple but useful scientific machine.** It is a ridge, or thread, wrapped around a bar or pole. It changes a small turning motion into a powerful pulling or lifting movement. Wood screws hold together furniture or shelves. A car jack lets you lift up a whole car.

▶ On a seesaw lever, the pivot is in the middle. Other levers have pivots at the end.

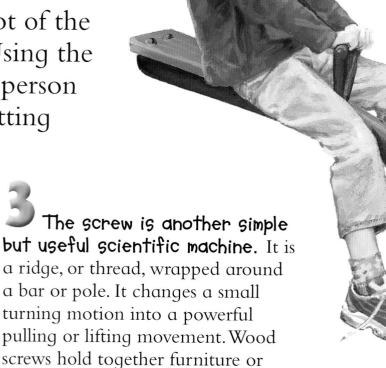

Axle

4 **Where would you be without wheels?** Not going very far. The wheel is a simple machine, a circular disk that turns around its center on a bar called an axle. Wheels carry heavy weights easily. There are giant wheels on big trucks and trains and small wheels on rollerblades.

▶ A car's rear wheels are turned by axles.

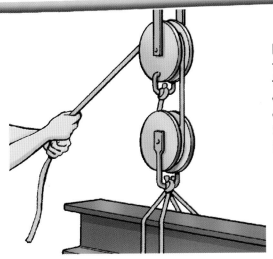

▲ Two pulleys together reduce the force needed to lift a heavy girder by one half.

▶ Gears change the turning direction of a force. They can slow it down or speed it up—and even convert it into a sliding force (rack and pinion).

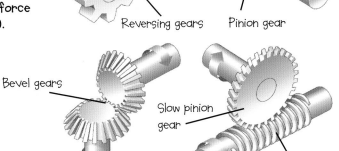

Reversing gears

Sliding rack

Pinion gear

Bevel gears

Slow pinion gear

Slow worm gear

5 A pulley turns around, like a wheel. It has a groove around its edge for a cable or rope. Lots of pulleys allow us to lift very heavy weights easily. The pulleys on a tower crane can lift huge steel girders to the top of a skyscraper.

6 Gears are like wheels, with pointed teeth around the edges. They change a fast, weak turning force into a slow, powerful one—or the other way around. On a bicycle, you can pedal up the steepest hill in bottom (lowest) gear, then speed down the other side in top (highest) gear.

Lever

Pivot

I DON'T BELIEVE IT!

A ramp is a simple machine called an inclined plane. It is easier to walk up a ramp than to jump straight to the top.

When science is hot!

7 Fire! Flames! Burning! Heat! The science of heat is important in all kinds of ways. Not only do we cook with heat, but we also warm our homes and heat water. Burning happens in all kinds of engines in cars, trucks, planes, and rockets. It is also used in factory processes, from making steel to shaping plastics.

8 **Heat moves by conduction.** A hot object will pass on, or transfer, some of its heat to a cooler one. Dip a metal spoon in a hot drink and the spoon handle soon warms up. Heat is conducted from the drink, through the metal.

▲ A firework burns suddenly as an explosive, with heat, light, and sound … BANG!

◄ Metal is a good conductor of heat. Put a teaspoon in a hot drink and feel how quickly it heats up.

9 **Heat moves by invisible "heat rays."** This is called thermal radiation and the rays are infrared waves. The Sun's warmth radiates through space as infrared waves, to reach Earth.

10 Burning, also called combustion, is a chemical process. Oxygen gas from the air joins to, or combines with, the substance being burned. The chemical change releases lots of heat, and usually light too. If this happens really fast, we call it an explosion.

11 Temperature is the amount of heat in a substance. It is usually measured in degrees Fahrenheit (°F) or Celsius (°C). Water freezes at 32°F (0°C), and boils at 212°F (100°C). We use a thermometer to take temperatures. Your body temperature is about 98.6°F (37°C).

▶ A thermometer may be filled with alcohol and red dye. As the temperature goes up, the liquid rises up its tube to show how hot it is. It sinks back down if the temperature falls.

CARRYING HEAT

You will need:

wooden ruler   metal spoon
plastic spatula   heatproof pitcher
frozen peas   some butter

Find a wooden ruler, a metal spoon, and a plastic spatula, all the same length. Fix a frozen pea to one end of each with butter. Put the other ends in a heatproof pitcher. Ask an adult to fill it with hot water. Heat is conducted from the water, up the object, to melt the butter. Which object is the best conductor?

12 Heat moves by convection, especially through liquids and gases. Some of the liquid or gas takes in heat, gets lighter and rises into cooler areas. Then other, cooler, liquid or gas moves in to do the same. You can see this as "wavy" hot air rising from a flame.

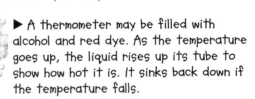

▶ See how hot air shimmers over a candle.

Engine power

13 **Imagine having to walk or run everywhere, instead of riding in a car.** Engines are machines that use fuel to do work for us and make life easier. Fuel is a substance that has chemical energy stored inside it. The energy is released as heat by burning or exploding the fuel in the engine.

Fan sucks air in

Air is squashed by turbines

Jet fuel is sprayed onto air, and small explosion happens

14 **Most cars have gasoline engines.** A mixture of air and gasoline is pushed into a hollow chamber called the cylinder. A spark from a spark plug makes it explode, which pushes a piston down inside the cylinder (see below). This movement is used by gears to turn the wheels. Most cars have four or six cylinders.

15 **A diesel engine works in a similar way, but without sparks.** The mixture of air and diesel is squashed so much in the cylinder that it becomes hot enough to explode. Diesel engines are used where lots of power is needed, in trucks, backhoes, tractors, and big trains.

▼ This shows the four-stroke cycle of a gasoline engine.

1. Air and gasoline mixture is sucked into the cylinder

2. The piston moves up and squeezes the mixture

3. A spark from the plug makes the mixture explode

4. The piston rises to push waste gases out of the cylinder

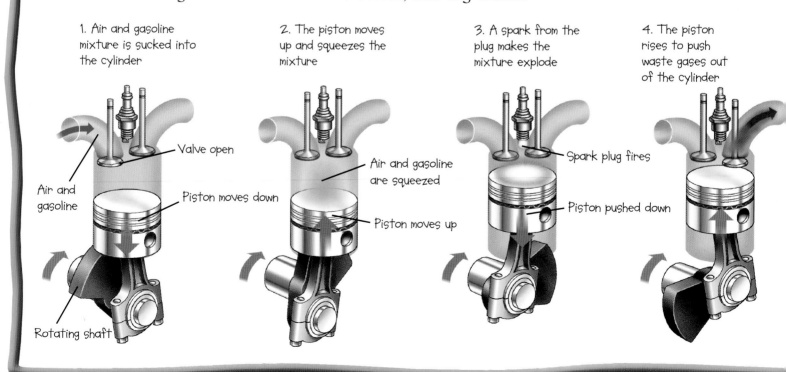

Valve open

Air and gasoline

Piston moves down

Rotating shaft

Air and gasoline are squeezed

Piston moves up

Spark plug fires

Piston pushed down

▼ Jet engines are very powerful. They use a mixture of air and fuel to push the plane forward at high speed.

Gases roar past exhaust turbines

Hot gases rush out of the engine

Afterburner adds more roaring gases

16 **A jet engine mixes air and kerosene and sets fire to it in one long, continuous, roaring explosion.** Incredibly hot gases blast out of the back of the engine. These push the engine forward—along with the plane.

17 **An electric motor passes electricity through coils of wire.** This makes the coils magnetic, and they push or pull against magnets around them. The push-pull makes the coils spin on their shaft (axle).

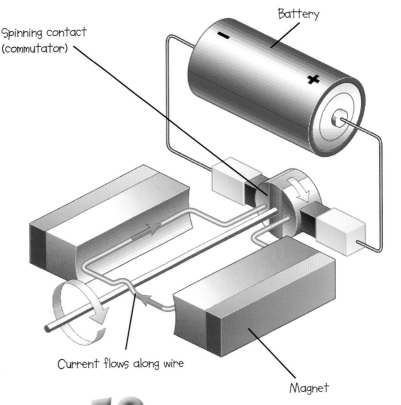

Battery

Spinning contact (commutator)

Current flows along wire

Magnet

QUIZ
Name the engine for each vehicle. Is it a jet engine, an electric motor, a gasoline engine, or a diesel one?
1. Cable car
2. Formula 1 racing car
3. Fork-lift truck
4. Land-speed record car

3. Diesel engine   4. Jet engine
2. Gasoline engine
1. Electric motor

18 **Engines that burn fuel give out gases and particles through their exhausts.** Some of these gases are harmful to the environment. The less we use engines, the better. Electric motors are quiet, efficient, and reliable, but they still need fuel—to make the electricity at the power station.

Science on the move

19 **Without science, we would have to walk, or ride a horse.** Luckily, scientists and engineers have developed many methods of transportation, most importantly, the car. Lots of people can travel together in a bus, train, plane, or ship. This uses less energy and resources, and makes less pollution.

Passenger terminal

Underground trains to take passengers to and from the terminal

20 **Science is used to stop criminals.** Science-based security measures include a "door frame" that detects metal objects like guns, and a scanner that sees inside bags. A sniffer-machine (or dog) can detect the smell of explosives or illegal drugs.

21 Jetways are extending walkways that stretch out like telescopic fingers, right to the plane's doors. Their supports move along on wheeled trolleys driven by electric motors.

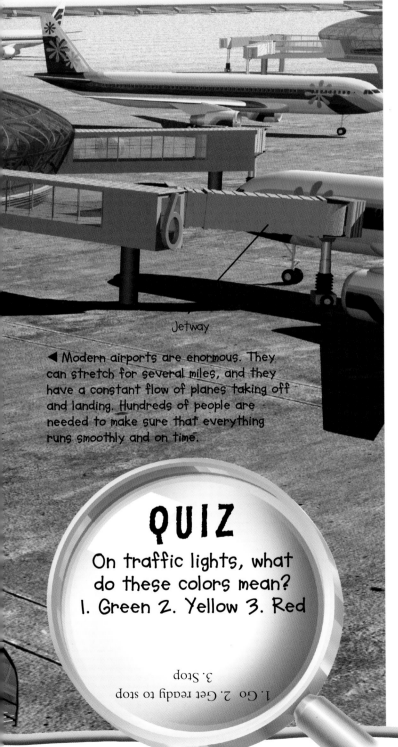

Jetway

◄ Modern airports are enormous. They can stretch for several miles, and they have a constant flow of planes taking off and landing. Hundreds of people are needed to make sure that everything runs smoothly and on time.

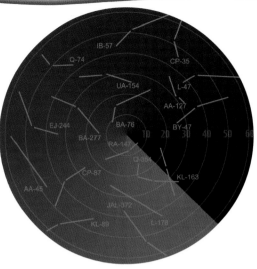

▶ The radar screen shows each aircraft as a blip, with its flight number or identity code.

22 Every method of transportation needs to be safe and on time. In the airport control tower, air traffic controllers track planes on radar screens. They talk to pilots by radio. Beacons send out radio signals, giving the direction and distance to the airport.

23 On the road, drivers obey traffic lights. On a train network, engineers obey similar signal lights of different colors, such as red for stop. Sensors by the track record each train passing and send the information by wires or radio to the control room. Each train's position is shown as a flashing light on a wall map.

▼ Train signals show just two colors—red for stop and green for go.

QUIZ
On traffic lights, what do these colors mean?
1. Green 2. Yellow 3. Red

1. Go 2. Get ready to stop 3. Stop

Noisy science

24 Listening to the radio or television, playing music, shouting at each other—they all depend on the science of sound—acoustics. Sounds are like invisible waves in the air. The peak (highest point) of the wave is where a region of air is squashed under high pressure. The trough (lowest point) of the wave is a region where air is expanded under low pressure.

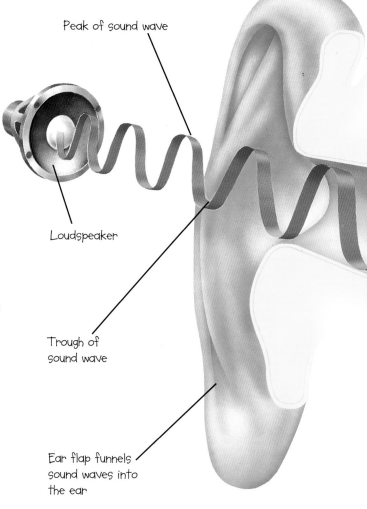

Peak of sound wave

Loudspeaker

Trough of sound wave

Ear flap funnels sound waves into the ear

▲ We cannot see sound waves but we can certainly hear them. They are ripples of high and low pressure in air.

25 Scientists measure the loudness or intensity of sound in decibels, dB. A very quiet sound like a ticking watch is 10dB. Ordinary speech is 50–60dB. Loud music is 90dB. A jet plane taking off is 120dB. Too much noise damages the ears.

Atomic explosion

Jet plane

Whisper   Express train

| O dB | 40 dB | 80 dB | 120 dB | 160 dB |

◄ The decibel scale measures the intensity, or energy, in sound.

26 Whether a sound is high or low is called its pitch, or frequency. It is measured in Hertz, Hz. A singing bird or whining motorcycle has a high pitch. A rumble of thunder or a massive truck has a low pitch. People can hear frequencies from 25 to 20,000Hz.

28

Sound waves travel about 1,100ft (330m) every second. This is still a million times slower than light waves. Sound waves also bounce off hard, flat surfaces. This is called reflection. The returning waves are heard as an echo.

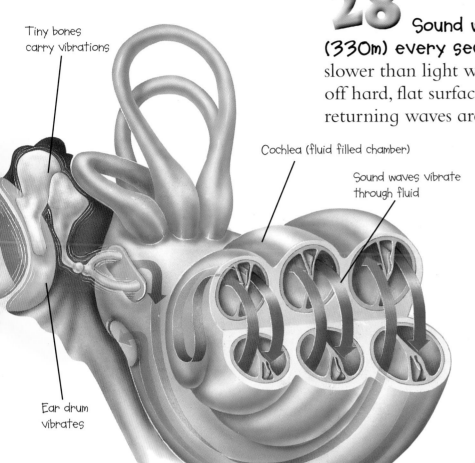

Tiny bones carry vibrations

Cochlea (fluid filled chamber)

Sound waves vibrate through fluid

Ear drum vibrates

29

Loudspeakers change electrical signals into sounds. The signals in the wire pass through a wire coil inside the speaker. This turns the coil into a magnet, which pushes and pulls against another magnet. The pushing and pulling make the cone vibrate, which sends sound waves into the air.

27

Sound waves spread out from a vibrating object that is moving rapidly to and fro. Stretch a rubber band between your fingers and twang it. As it vibrates, it makes a sound. When you speak, vocal cords in your neck vibrate. You can feel them through your skin.

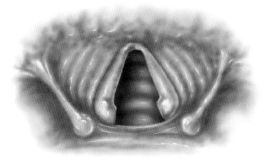

◀ The vocal cords are tough flaps in your voicebox, in your neck.

BOX GUITAR

You will need:

shoebox   rubber band

split pins   cardboard

Cut a hole about 4in (10cm) across on one side of an empy shoebox. Push split pins through either side of the hole, and stretch a rubber band between them. Pluck the band. Hear how the air and box vibrate. Cover the hole with card. Is the "guitar" as loud?

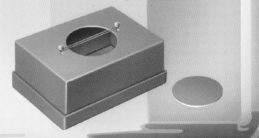

Look out—light's about!

30 Almost everything you do depends on light and the science of light, which is called optics. Light is a form of energy that you can see. Light waves are made of electricity and magnetism—and they are tiny. About 2,000 of them laid end to end would stretch across this period.

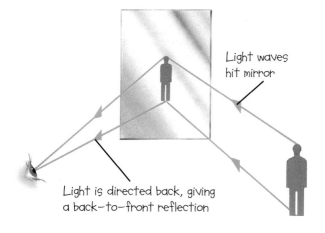

Light waves hit mirror

Light is directed back, giving a back-to-front reflection

▲ Light waves bounce off a mirror.

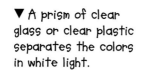

▼ A prism of clear glass or clear plastic separates the colors in white light.

32 Like sound, light bounces off surfaces that are **very smooth.** This is called reflection. A mirror is smooth, hard and flat. When you look at it, you see your reflection.

31 Ordinary light from the Sun or from a light bulb is called **white light.** But when white light passes through a prism, a triangular block of clear glass, it splits into seven colors. These colors are known as the spectrum. Each color has a different length of wave. A rainbow is made by raindrops, which work like millions of tiny prisms to split up sunlight.

33 Light passes through certain materials, like clear glass and plastic. Materials that let light pass through, to give a clear view, are transparent. Those that do not allow light through, like wood and metal, are opaque.

► Glass and water bend, or refract, light waves. This makes a drinking straw look bent where it goes behind the glass and then into the water.

34 Mirrors and lenses are important parts of many optical (light—using) gadgets. They are found in cameras, binoculars, microscopes, telescopes, and lasers. Without them, we would have no close-up photographs of tiny microchips or insects or giant planets—in fact, no photos at all.

35 Light does not usually go straight through glass. It bends slightly where it goes into the glass, then bends back as it comes out. This is called refraction. A lens is a curved piece of glass or plastic that bends light to make things look bigger, smaller or clearer. Eyeglasses and contact lenses bend light to help people see more clearly.

▼ A concave lens, which is thin in the middle, makes things look smaller.

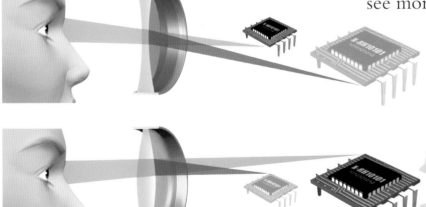

▲ A convex lens, which bulges in the middle, makes things look larger.

I DON'T BELIEVE IT!
Light is the fastest thing in the Universe. It travels 186,000mi (300,000km) per second. That's seven times around the world in less than one second!

The power of lasers

36 **Laser light is a special kind of light.** Like ordinary light, it is made of waves, but there are three main differences. First, ordinary white light is a mixture of colors. Laser light is just one pure color. Second, ordinary light waves have peaks (highs) and troughs (lows), which do not line up with each other—laser light waves line up perfectly. Third, an ordinary light beam spreads and fades. A laser beam does not. It can travel for thousands of miles as a strong, straight beam.

37 **To make a laser beam, energy is fed in short bursts into a substance called the active medium.** The energy might be electricity, heat, or ordinary light. In a red ruby laser, the active medium is a rod of ruby crystal. A strong lamp makes the tiny particles in the crystal vibrate. They give off energy, which bounces to and fro inside the crystal, off the mirrors at each end. Eventually, the rays vibrate with each other and they are all the same length. The energy becomes so strong that it bursts through a mirror at the end of the crystal.

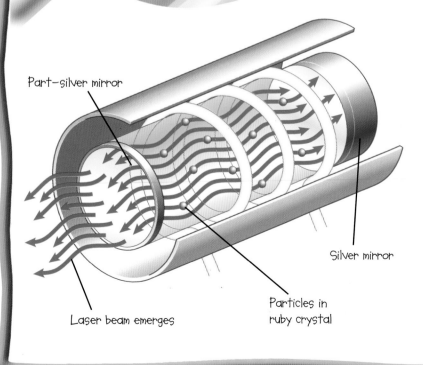

Part-silver mirror

Silver mirror

Particles in ruby crystal

Laser beam emerges

◀ Waves of light build up and bounce to and fro inside a laser, then emerge at one end.

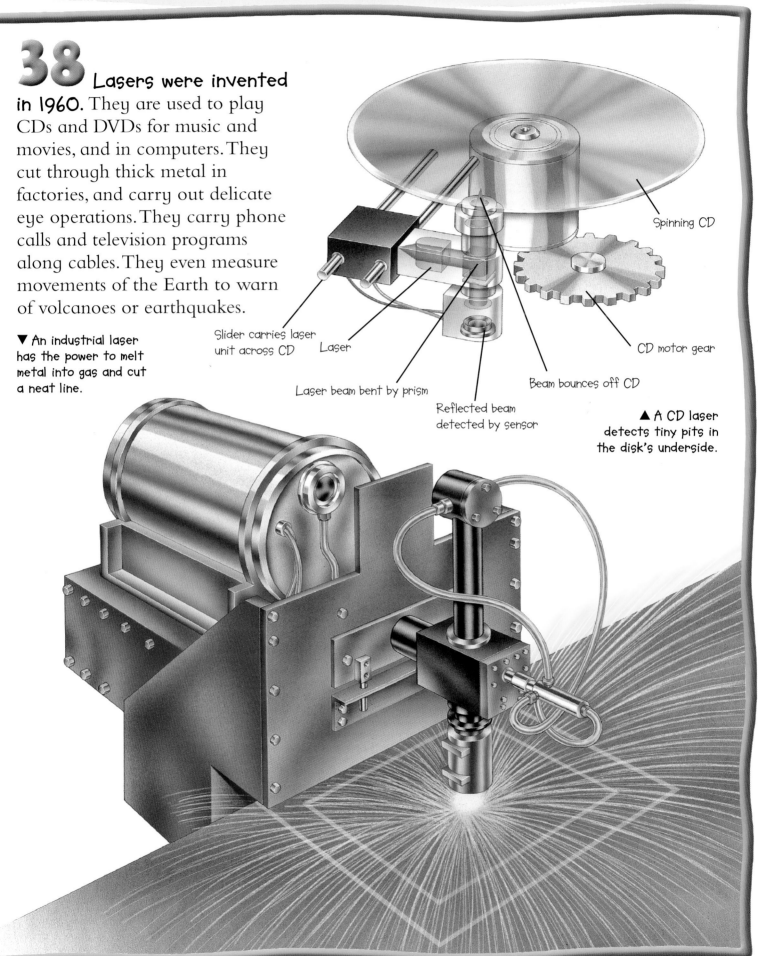

38 Lasers were invented in 1960.

They are used to play CDs and DVDs for music and movies, and in computers. They cut through thick metal in factories, and carry out delicate eye operations. They carry phone calls and television programs along cables. They even measure movements of the Earth to warn of volcanoes or earthquakes.

▼ An industrial laser has the power to melt metal into gas and cut a neat line.

Spinning CD

Slider carries laser unit across CD

Laser

CD motor gear

Laser beam bent by prism

Beam bounces off CD

Reflected beam detected by sensor

▲ A CD laser detects tiny pits in the disk's underside.

Mystery magnets

39 Without magnets there would be no electric motors, computers, or loudspeakers. Magnetism is an invisible force to do with atoms—tiny particles that make up everything. Atoms are made of even smaller particles, including electrons. Magnetism is linked to the way that these line up and move. Most magnetic substances contain iron. As iron makes up a big part of the metallic substance steel, steel is also magnetic.

40 A magnet is a lump of iron or steel that has all its electrons and atoms lined up. This means that their magnetic forces all add up. The force surrounds the magnet, in a region called the magnetic field. This is strongest at the two parts of the magnet called the poles. In a bar or horseshoe magnet, the poles are at the ends.

▶ An electromagnet attracts the body of a car, which is made of iron—based steel.

▼ The field around a magnet affects objects that contain iron.

Magnetic field

Magnetic lines of force

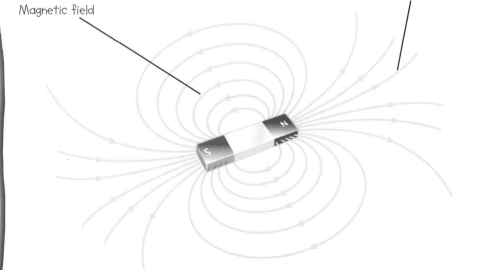

41 **When electricity flows through a wire, it makes a weak magnetic field around it.** If the wire is wrapped into a coil, the magnetism becomes stronger. This is called an electromagnet. Its magnetic force is the same as an ordinary magnet, but when the electricity goes off, the magnetism does too. Some electromagnets are so strong, they can lift whole cars.

42 **A magnet has two different poles—north and south.** A north pole repels (pushes away) the north pole of another magnet. Two south poles also repel each other. But a north pole and a south pole attract (pull together). Both magnetic poles attract any substance containing iron, like a nail or a screw.

VA 2314

QUIZ
Which of these substances or objects is magnetic?
1. Metal spoon 2. Plastic spoon
3. Pencil 4. Drinks can
5. Food can 6. Screwdriver
7. Cooking foil

1.Yes 2.No 3.No
4.No 5.Yes 6.Yes 7.No

Electric sparks!

43 **Flick the switch and things happen.** The television goes off, the computer comes on, lights shine, and music plays. Electricity is our favorite form of energy. We send it along wires and plug hundreds of machines into it. Imagine no washing machine, no electric light, and no vacuum cleaner!

► Electricity is bits of atoms moving along a wire.

Atom

Electron

44 **Electricity depends on electrons, tiny parts of atoms.** In certain substances, when electrons are "pushed," they hop from one atom to the next. When billions do this every second, electricity flows. The "push" is from a battery or the generator at a power station. Electricity only flows if it can go in a complete loop or circuit. Break the circuit and the flow stops.

45 **A battery makes electricity from chemicals.** Two different chemicals next to each other, such as an acid and a metal, swap electrons and get the flow going. Electricity's pushing strength is measured in volts. Most batteries are about 1.5, 3, 6, or 9 volts, with 12 volts in cars.

46 **Electricity flows easily through some substances, including water and metals.** These are electrical conductors. Other substances do not allow electricity to flow. They are insulators. Insulators include wood, plastic, glass, cardboard and ceramics. Metal wires and cables have coverings of plastic, to stop the electricity leaking away.

Positive contact

◄ A battery has a chemical paste inside a metal casing.

Negative contact on base

47

Electricity from power stations is carried along cables on high pylons, or buried underground. This is known as the distribution grid. At thousands of volts, this electricity is extremely dangerous. For use in the home, it is changed to 110 volts. But it can still easily kill a person.

▼ A power station makes enough electricity for thousands of homes.

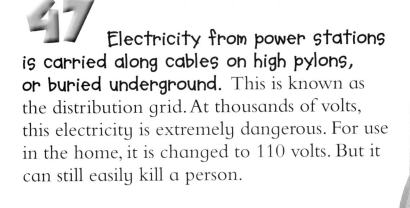

◀ High pylons hold electric cables safely above ground.

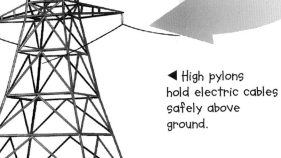

48

Mains electricity is made at a power station. A fuel such as coal or oil is burned to heat water into high-pressure steam. The steam pushes past the blades of a turbine and makes them spin. The spinning motion turns coils of wire near powerful magnets, and this makes electricity flow in the coils.

MAKE A CIRCUIT
You will need:

lightbulb   battery
some wire   plastic ruler
metal spoon   dry cardboard

Join a bulb to a battery with pieces of wire, as shown. Electricity flows around the circuit and lights the bulb. Make a gap in the circuit and put various objects into it, to see if they allow electricity to flow again. Try a plastic ruler, a metal spoon and some cardboard.

Making sounds and pictures

49 **The air is full of waves we cannot see or hear, unless we have the right machine.** Radio waves are a form of electrical and magnetic energy, just like heat and light waves, microwaves and X-rays. All of these are called electromagnetic waves and they travel at an equal speed—the speed of light.

50 **Radio waves are used for both radio and television.** They travel vast distances. Long waves curve around the Earth's surface. Short waves bounce between the Earth and the sky.

▼ All these waves are the same form of energy. They all differ in length.

▲ A radio set picks up radio waves using its long antenna.

51 **Radio waves carry their information by being altered, or modulated, in a certain pattern.** The height of a wave is called its amplitude. If this is altered, it is known as AM (amplitude modulation). Look for AM on the radio display.

52 **The number of waves per second is called the frequency.** If this is altered, it is known as FM (frequency modulation). FM radio is clearer than AM, and less affected by weather and thunderstorms.

Long radio waves

Shorter radio waves (TV)

Microwaves

Light waves

X-rays

Short X-rays

Gamma rays

53

Radio and TV programs may be sent out as radio waves from a tall tower on the ground. The tower is called a transmitter. Sometimes waves may be broadcast (sent) by a satellite in space. Or the programs may not even arrive as radio waves. They can come as flashes of laser light, as cable TV and radio.

▶ A dish-shaped receiver picks up radio waves for TV channels.

54

Inside a TV set, the pattern of radio waves is changed into electrical signals. Some go to the loudspeaker to make the sounds. Others go to the screen to make the pictures. Inside most televisions, the screen is at the front of a glass container called a tube. At the back of the tube are electron guns. These fire streams of electrons. The inside of the screen is coated with thousands of tiny colored dots called phosphors. When electrons hit the dots, they glow and make the picture.

▼ A TV screen's three colors of patches or dots combine to make up the other colors.

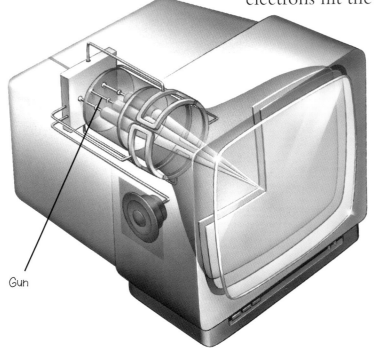

Gun

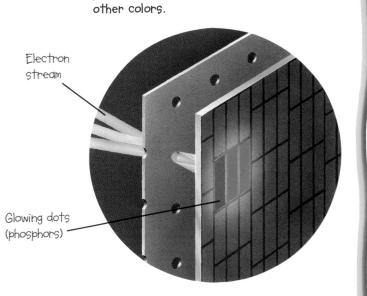

Electron stream

Glowing dots (phosphors)

Compu-science

55 **Computers are amazing machines.** But they have to be told exactly what to do. So we put in instructions and information, by various means. These include typing on a keyboard, inserting a disk, using a joystick or games board, or linking up a camera, scanner or another computer.

CD or DVD drive (reader)

Main computer case

Microchips on circuit board

Floppy disk drive

56 **Most computers are controlled by instructions from a keyboard and a mouse.** The mouse moves a pointer around on the screen and its click buttons select choices from lists called menus.

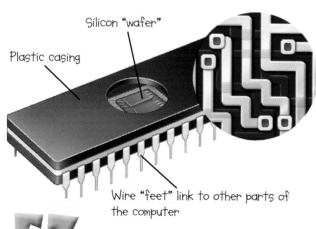

Silicon "wafer"

Plastic casing

Wire "feet" link to other parts of the computer

◄ This close up of a slice of silicon "wafer" shows the tiny parts which receive and send information in a computer.

57 **Some computers are controlled by talking to them!** They pick up the sounds using a microphone. This is VR, or voice recognition technology.

58 **The "main brain" of a computer is its Central Processing Unit.** It is usually a microchip—millions of electronic parts on a chip of silicon, hardly larger than a fingernail. It receives information and instructions from other microchips, carries out the work, and sends back the results.

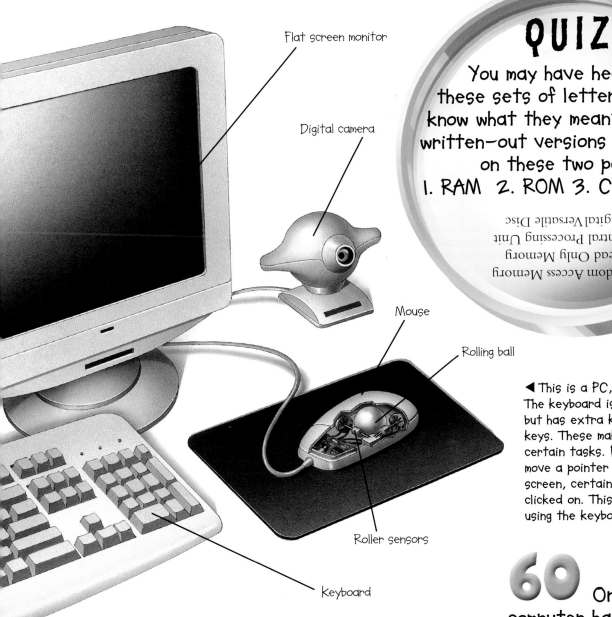

Flat screen monitor

Digital camera

Mouse

Rolling ball

Roller sensors

Keyboard

QUIZ

You may have heard of these sets of letters. Do you know what they mean? Their full written-out versions are all here on these two pages.

1. RAM 2. ROM 3. CPU 4. DVD

1. Random Access Memory
2. Read Only Memory
3. Central Processing Unit
4. Digital Versatile Disc

◄ This is a PC, or personal computer. The keyboard is like a typewriter, but has extra keys called function keys. These make the computer do certain tasks. By using the mouse to move a pointer (cursor) around the screen, certain instructions can be clicked on. This can be quicker than using the keyboard.

60 Once the computer has done its task, it feeds out the results. These usually go to a screen called a monitor, where we see them. But they can also go to a printer, a loudspeaker or even a robot arm. Or they can be stored on a disk such as a magnetic disk, compact disk or Digital Versatile Disc (DVD).

59 Information and instructions are contained in the computer in memory microchips. There are two kinds. Random Access Memory is like a jotting pad. It keeps changing as the computer carries out its tasks. Read Only Memory is like an instruction book. It usually contains the instructions for how the computer starts up and how all the microchips work together.

Web around the world

61 The world is at your fingertips—if you are on the Internet. The "Net" is one of the most amazing results of modern-day science. It is a worldwide network of computers, linked like one huge electrical spiderweb.

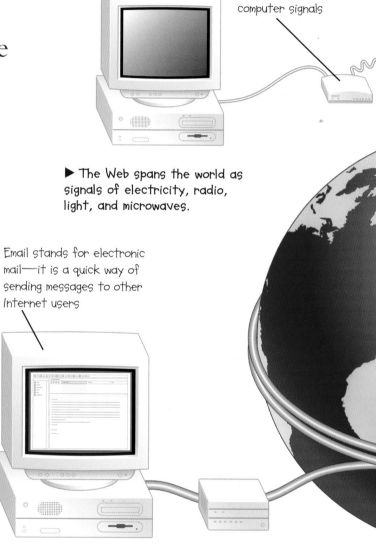

A modem changes telephone signals to computer signals

► The Web spans the world as signals of electricity, radio, light, and microwaves.

Email stands for electronic mail—it is a quick way of sending messages to other Internet users

▲ As cellular phones get smaller, they can also connect to the Net using their radio link.

62 Signals travel from computer to computer in many ways. These include electricity along telephone wires, flashes of laser light along fiber-optic cables or radio waves between tall towers. Information is changed from one form to another in a split second. It can also travel between computers on different sides of the world in less than a second using satellite links.

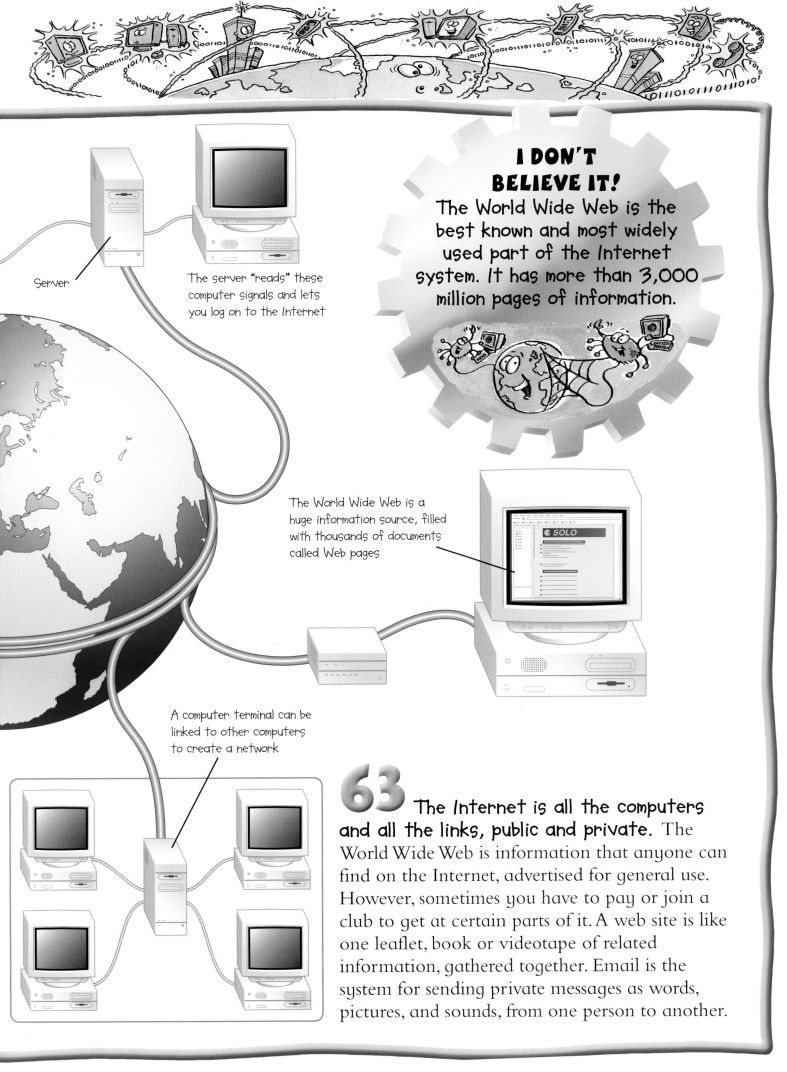

Server

The server "reads" these computer signals and lets you log on to the Internet

The World Wide Web is a huge information source, filled with thousands of documents called Web pages

SOLO

A computer terminal can be linked to other computers to create a network

63 The Internet is all the computers and all the links, public and private. The World Wide Web is information that anyone can find on the Internet, advertised for general use. However, sometimes you have to pay or join a club to get at certain parts of it. A web site is like one leaflet, book or videotape of related information, gathered together. Email is the system for sending private messages as words, pictures, and sounds, from one person to another.

What's it made of?

64 You would not make a bridge out of straw, or a cup out of thin paper! Choosing the right substance or material for the job is part of materials science. All the substances in the world can be divided into several groups. The biggest group is metals such as iron, copper, silver, and gold. Most metals are strong, hard and shiny, and carry heat and electricity well. They are used where materials must be tough and long-lasting.

65 Plastics are made mainly from the substances in petroleum (crude oil). There are so many kinds—some are hard and brittle but others are soft and bendy. They are usually long-lasting, not affected by weather or damp, and they resist heat and electricity.

▼ A racing car has thousands of parts made from hundreds of materials. Each is suited to certain conditions such as stress, temperature, and vibrations.

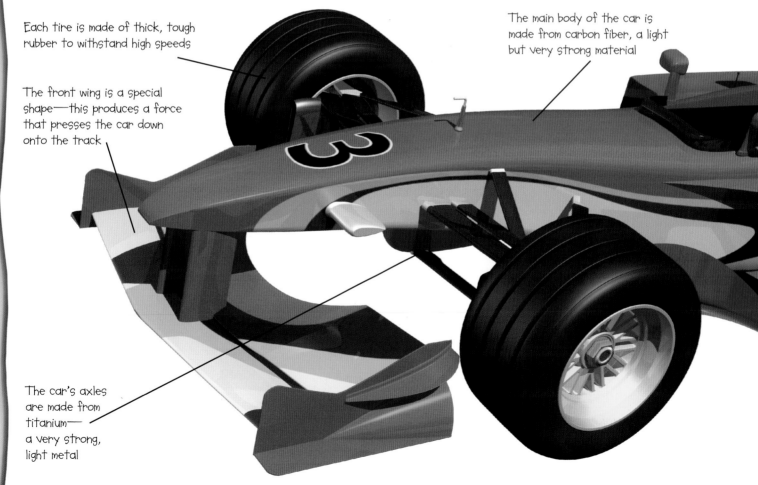

Each tire is made of thick, tough rubber to withstand high speeds

The front wing is a special shape—this produces a force that presses the car down onto the track

The car's axles are made from titanium—a very strong, light metal

The main body of the car is made from carbon fiber, a light but very strong material

66 Ceramics are materials based on clay or other substances dug from the Earth. They can be shaped and dried, like a clay bowl. Or they can be fired—baked in a hot oven called a kiln. This makes them hard and long-lasting, but brittle and prone to cracks. Ceramics resist heat and electricity very well.

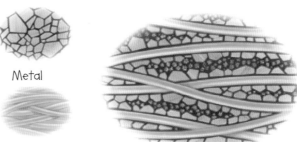

Metal

Fiber

Ceramic

◀ Metal, fiber and ceramic can combine to make a composite material (above). The way all of these ingredients are arranged can affect the composite's strength.

67 Glass is produced from the raw substances limestone and sand. When heated at a high temperature, these substances become a clear, gooey liquid, which sets hard as it cools. Its great advantage is that you can see through it.

68 Composites are mixtures or combinations of different materials. For example, glass strands are coated with plastic to make GRP—glass-reinforced plastic. This composite has the advantages of both materials.

Rear wing

The engine can produce about 10 times as much power as an ordinary car—but it needs to be as light as possible

MAKE YOUR OWN COMPOSITE
You will need:

flour   newspaper strips
water   balloon   pin

You can make a composite called pâpier-maché from flour, newspaper, and water. Tear newspaper into strips. Mix flour and water into a paste. Dip each strip in the paste and place it around a blown-up balloon. Cover the balloon and allow it to dry. Pop the balloon with a pin, and the composite should stay in shape.

The world of chemicals

69 **The world is made of chemical substances.** Some are completely pure. Others are mixtures of substances—such as petroleum (crude oil). Petroleum provides us with thousands of different chemicals and materials, such as plastics, paints, soaps, and fuel. It is one of the most useful, and valuable, substances in the world.

Fumes and vapors condense into liquid

▶ The huge tower (fractionating column) of an oil refinery may be 165ft (50m) high.

70 **In an oil refinery, crude oil is heated in a huge tower.** Some of its different substances turn into fumes (vapors) and rise up the tower. The fumes turn back into liquids at different heights inside the tower, due to the different temperatures at each level. We get gasoline in this way. Remaining at the bottom are thick, gooey tars, asphalts, and bitumens —which are used to make road surfaces.

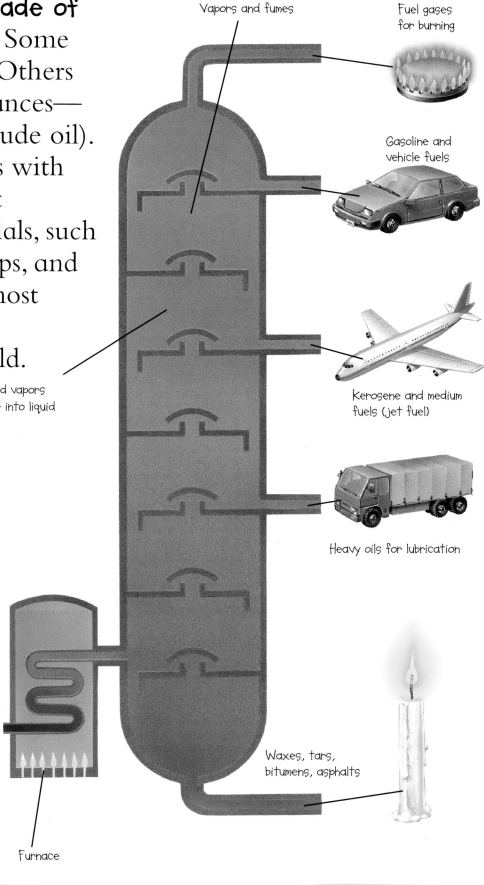

Vapors and fumes

Fuel gases for burning

Gasoline and vehicle fuels

Kerosene and medium fuels (jet fuel)

Heavy oils for lubrication

Waxes, tars, bitumens, asphalts

Furnace

71

One group of chemicals is called acids. They vary in strength from very weak citric acid, which gives the sharp taste to fruits such as lemons, to extremely strong and dangerous sulfuric acid in a car battery. Powerful acids burn and corrode, or eat away, substances. Some even corrode glass or steel.

Acidic substance

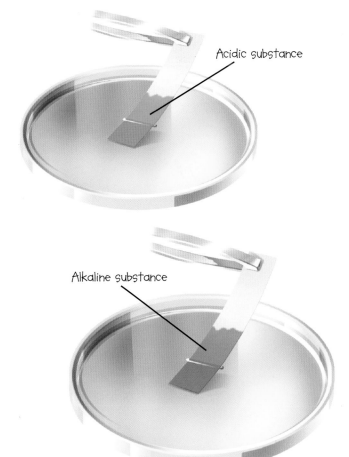

Alkaline substance

72

Another group of chemicals is bases. They vary in strength from weak alkaloids, which give the bitter taste to coffee beans, to strong and dangerous bases in drain cleaners and industrial polishes. Bases feel soapy or slimy and, like acids, they can burn or corrode.

▼ Indicator paper changes color when it touches different substances. Acids turn it red, alkalis make it bluish-purple. The deeper the color, the stronger the acid or base.

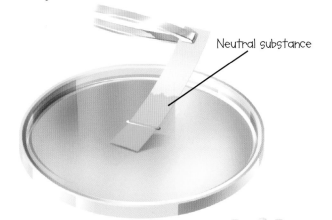

Neutral substance

73

Acids and bases are "opposite" types of chemicals. When they meet, they undergo changes called a chemical reaction. The result is usually a third type of chemical, called a salt. The common salt we use for cooking is one example. Its chemical name is sodium chloride.

FROTHY FUN
You will need:

some vinegar   washing soda

Create a chemical reaction by adding a few drops of vinegar to a spoonful of washing soda in a saucer. The vinegar is an acid, the soda is a base. The two react by frothing and giving off bubbles of carbon dioxide gas. What is left is a salt (but not to be eaten).

Pure science

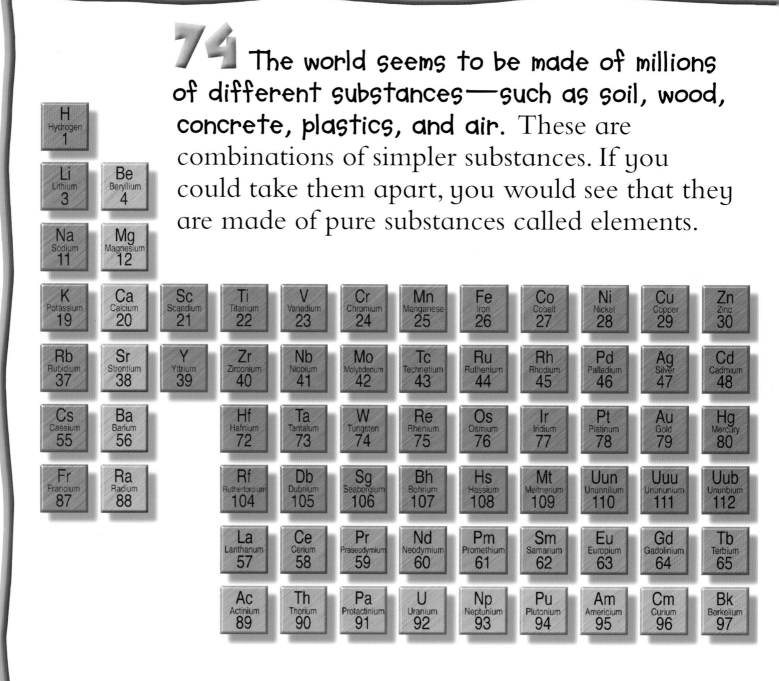

74 The world seems to be made of millions of different substances—such as soil, wood, concrete, plastics, and air. These are combinations of simpler substances. If you could take them apart, you would see that they are made of pure substances called elements.

75 Hydrogen is the simplest element. This means it has the smallest atoms. It is a very light gas, which floats upward in air. Hydrogen was once used to fill giant airships. But there was a problem —hydrogen catches fire easily and explodes. In fact, stars are made mainly of burning hydrogen, which is why they are so hot and bright.

76 About 90 elements are found naturally on and in the Earth. In an element, all of its particles, called atoms, are exactly the same as each other. Just as important, they are all different from the atoms of any other element.

▼ The elements can be arranged in a table. Each has a letter, like C for carbon. It also has a number showing how big or heavy its atoms are compared to those of other elements.

					He Helium 2
B Boron 5	C Carbon 6	N Nitrogen 7	O Oxygen 8	F Flourine 9	Ne Neon 10
Al Alumnium 13	Si Silicon 14	P Phosphorus 15	S Sulphur 16	Cl Chlorine 17	Ar Argon 18
Ga Gallium 31	Ge Germanium 32	As Arsenic 33	Se Selenium 34	Br Bromine 35	Kr Krypton 36
In Indium 49	Sn Tin 50	Sb Antimony 51	Te Tellurium 52	I Iodine 53	Xe Xenon 54
Tl Thalium 81	Pb Lead 82	Bi Bismuth 83	Po Polonium 84	At Astatine 85	Rn Radon 86

Dy Dysprosium 66	Ho Holmium 67	Er Erbium 68	Tm Thulium 69	Yb Ytterbium 70	Lu Lutetium 71
Cf Californium 98	Es Einsteinium 99	Fm Fermium 100	Md Mendelevium 101	No Nobelium 102	Lr Lawrencium 103

QUIZ

1. Where does gasoline come from?
2. What usually happens when you mix an acid and a base?
3. Which element makes up stars?
4. What do diamonds and coal have in common?

1. Petroleum 2. They react to form a salt 3. Hydrogen 4. They are both made of pure carbon

78 **Uranium is a heavy and dangerous element.** It gives off harmful rays and tiny particles, called radioactivity. These can cause sickness, burns, and diseases such as cancer. Radioactivity is a type of energy and, under careful control, it may be used as fuel in nuclear power stations.

79 **Aluminum is an element that is a metal, and it is one of the most useful in modern life.** It is light and strong, it does not rust, and it is resistant to corrosion. Saucepans, drinks cans, cooking foil, and jet planes are made mainly of aluminum.

77 **Carbon is a very important element in living things—including our own bodies.** It joins easily with atoms of other elements to make large groups of atoms called molecules. When it is pure, carbon can be two different forms. These are soft, powdery soot, and hard, glittering diamond. The form depends on how the carbon atoms join to each other.

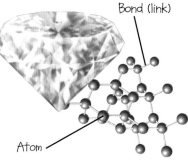

Bond (link)

▶ Carbon can be hard diamond or soft soot, which is made of sheets of joined atoms.

Atom

Small science

80 **Many pages in this book mention atoms.** They are the smallest bits of a substance. They are so tiny, even a billion atoms would be too small to see. But scientists have carried out experiments to find out what's inside an atom. The answer is—even smaller bits. These are subatomic particles, and there are three main kinds.

81 **At the center of each atom is a blob called the nucleus.** It contains an equal number of two kinds of subatomic particles. These are protons and neutrons. The proton is like the north pole of a magnet. It is positive, or plus. The neutron is not. It is neither positive or negative.

I DON'T BELIEVE IT!

One hundred years ago, people thought the electrons were spread out in an atom, like the raisins in a raisin pudding.

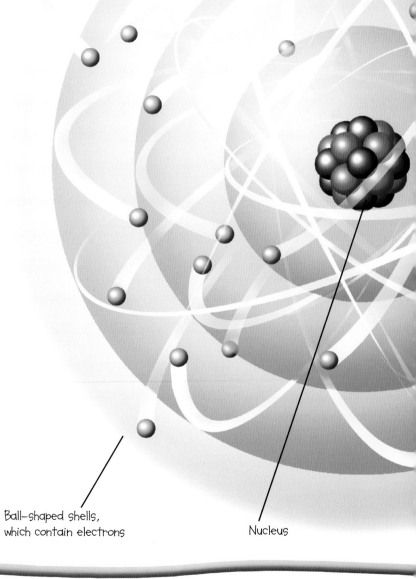

Electron

Ball-shaped shells, which contain electrons

Nucleus

82 Atoms of the various elements have different numbers of protons and neutrons. An atom of hydrogen has just one proton. An atom of helium, the gas put in party balloons to make them float, has one proton and one neutron. An atom of the heavy metal called lead has 82 protons and neutrons.

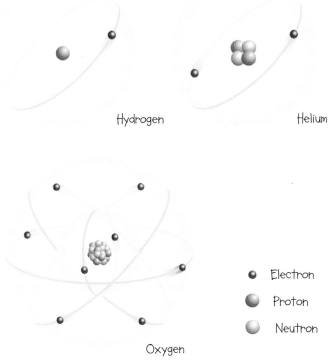

Hydrogen

Helium

Oxygen

● Electron

● Proton

● Neutron

▶ The bits inside an atom give each substance its features, from exploding hydrogen to life-giving oxygen.

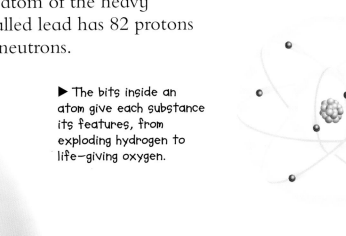

Movement of electrons

◀ Uranium is a tough, heavy, dangerous metal. Each atom of uranium has 92 electrons whizzing around its nucleus.

83 Around the center of each atom are subatomic particles called electrons. They whiz all around the nucleus. In the same way that a proton in the nucleus is positive or plus, an electron is negative or minus. The number of protons and neutrons is usually the same, so the plus and minus numbers are the same. (Electrons are the bits that jump from atom to atom when electricity flows.)

84 It is hard to imagine that atoms are so small. A grain of sand, smaller than this o, contains at least 100 billion billion atoms. If you could make each atom as big as a pinhead, the grain of sand would be more than 1 mile (1.6km) high!

Scientists at work

85 **There are thousands of different jobs and careers in science.** Scientists work in laboratories, factories, offices, mines, steelworks, nature parks, and almost everywhere else. They find new knowledge and make discoveries using a process called the scientific method.

86 **First comes an idea, called a theory or hypothesis.** This asks or predicts what will happen in a certain situation. Scientists continually come up with new ideas and theories to test. One very simple theory is—if I throw a ball up in the air, will it come back down?

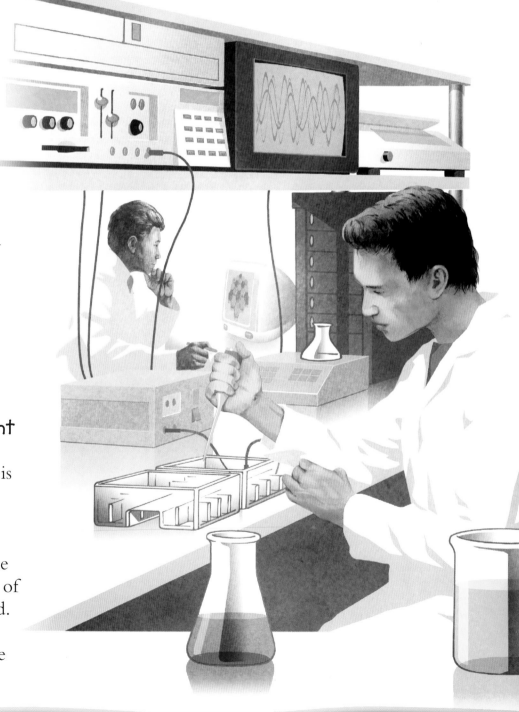

87 **The scientist carries out an experiment or test, to check what happens.** The experiment is carefully designed and controlled, so that it will reveal useful results. Any changes are carried out one at a time, so that the effect of each change can be studied. The experiment for our simple theory is—throw the ball up in the air.

88 Measuring and recording are very important as part of the experiment. All the changes are measured, written down, and perhaps photographed or filmed as well.

▼ Scientists carrying out research in a laboratory gather information and record all of their findings.

89 The results are what happens during and at the end of the experiment. They are studied, perhaps by drawing graphs and making tables. You can probably guess the result of our experiment—the ball falls back down.

90 At the end of this scientific process, the scientist thinks of reasons or conclusions about why certain things happened. The conclusion for our experiment is—something pulls the ball back down. But science never stands still. There are always new theories, experiments and results. This is how science progresses, with more discoveries and inventions every year.

QUIZ

Put these activities in the correct order, so that a scientist can carry out the scientific method.
1. Results 2. Experiment
3. Conclusions 4. Theory
5. Measurements

4,2,5,1,3

41

Science in nature

91 **Science and its effects are found all over the natural world.** Scientists study animals, plants, rocks, and soil. They want to understand nature, and find out how science and its technology affect wildlife.

92 **One of the most complicated types of science is ecology.** Ecologists try to understand how the natural world links together. They study how animals and plants live, what animals eat, and why plants grow better in some soils than others. They count the numbers of animals and plants and may trap animals briefly to study them, or follow the growth of trees in a wood. When the balance of nature is damaged, ecologists can help to find out why.

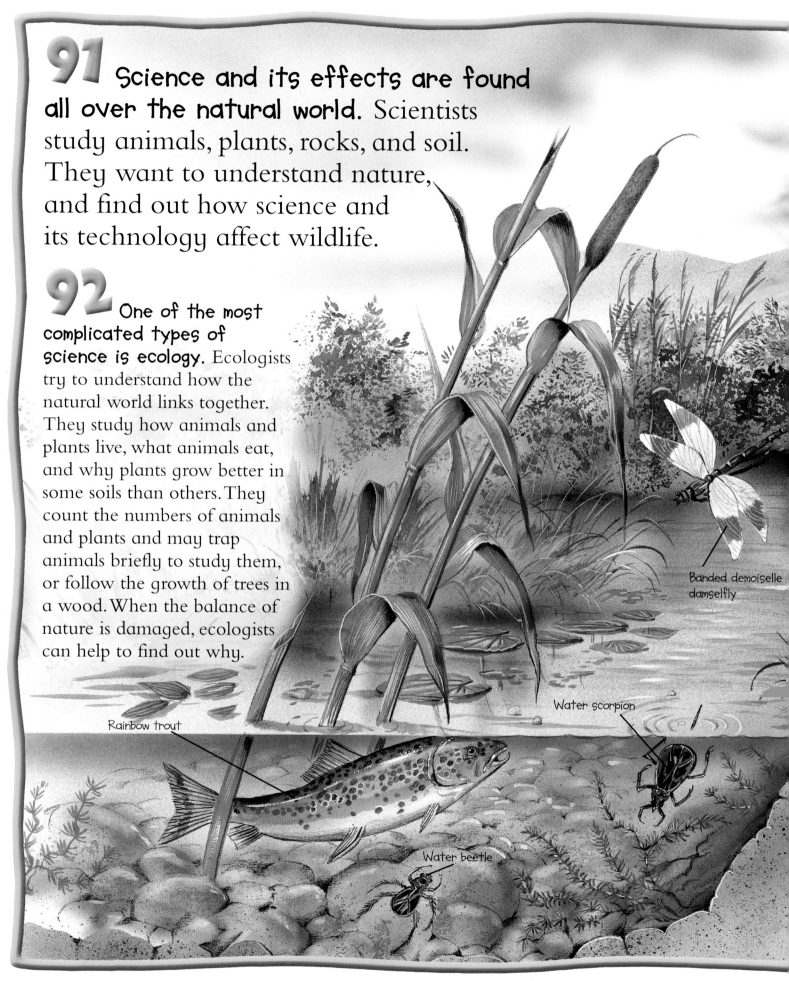

Banded demoiselle damselfly

Water scorpion

Rainbow trout

Water beetle

▼ One of the most important jobs in science is to study damage and pollution in the natural world. Almost everything we make or do affects wild places with their animals and plants. Factories, power stations, and roads crammed with vehicles are especially harmful, as chemicals spread in the air and seep into soil and water.

Reedmace

Power station

Heron

Otter

Warbler

93 Ecologists use many forms of high-tech science in their studies. They may fit an animal with a radio collar so that its movements can be tracked. Special cameras see in the dark and show how night hunters catch their prey. Radar used to detect planes can also follow flocks of birds. The sonar (echo-sounding) equipment of boats can track schools of fish or whales.

I DON'T BELIEVE IT!

Science explains how animals such as birds or whales find their way across the world. Some detect the Earth's magnetism, and which way is north or south. Others follow changes in gravity, the force that pulls everything to the Earth's surface.

Body science

94 One of the biggest areas of science is medicine. Medical scientists work to produce better drugs, more spare parts for the body, and more machines for use by doctors. They also carry out scientific research to find out how people can stay healthy and prevent disease.

▼ As a runner gets tired, his heart pumps harder. Its beats can be detected and shown on an ECG machine.

ECG machine showing display

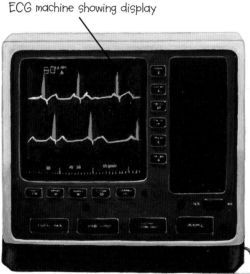

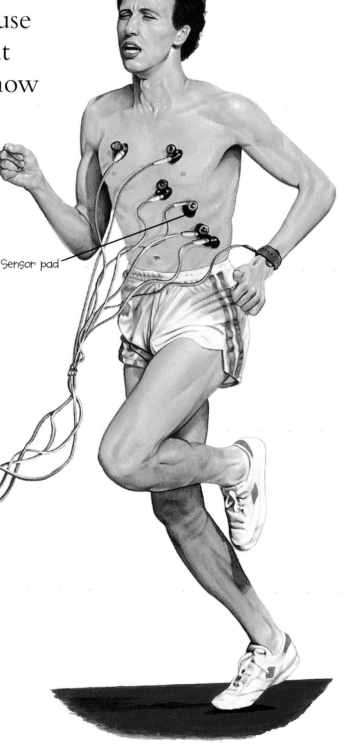

Sensor pad

95 As parts of the body work, such as the muscles and nerves, they produce tiny pulses of electricity. Pads on the skin pick up these pulses, which are displayed as a wavy line on a screen or paper strip. The ECG (electrocardiograph) machine shows the heart beating. The EEG (electroencephalograph) shows nerve signals flashing around the brain.

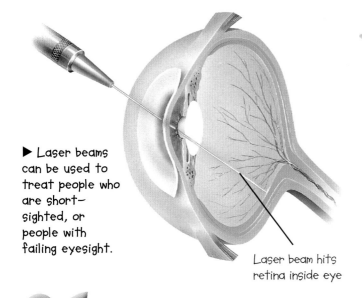

► Laser beams can be used to treat people who are short-sighted, or people with failing eyesight.

Laser beam hits retina inside eye

96 Laser beams are ideal for delicate operations, or surgery, on body parts such as the eye. The beam makes very small, precise cuts. It can be shone into the eye and made most focused, or strongest, inside. So it can make a cut deep within the eye, without any harm to the outer parts.

▼ An endoscope is inserted into the body to give a doctor a picture on screen. The treatment can be given immediately.

MAKE A PULSE MACHINE
You will need:

modeling clay   drinking straw

Find your pulse by feeling your wrist, just below the base of your thumb, with a finger of the other hand. Place some modeling clay on this area, and stick a drinking straw into it. Watch the straw twitch with each heartbeat. Now you can see and feel your pulse. Check your pulse rate by counting the number of heartbeats in one minute.

97 The endoscope is like a flexible telescope made of fiber-optic strands. This is pushed into a body opening such as the mouth, or through a small cut, to see inside. The surgeon looks into the other end of the endoscope, or at a picture on a screen.

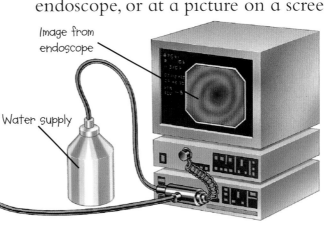

Endoscope tube

Image from endoscope

Water supply

45

Science in the future

98 Many modern machines and processes can cause damage to our environment and our health. The damage includes acid rain, destruction of the ozone layer, and the greenhouse effect, leading to climate change and global warming. Science can help to find solutions. New filters and chemicals called catalysts can reduce dangerous fumes from vehicle exhausts and power stations, and in the chemicals in factory waste pipes.

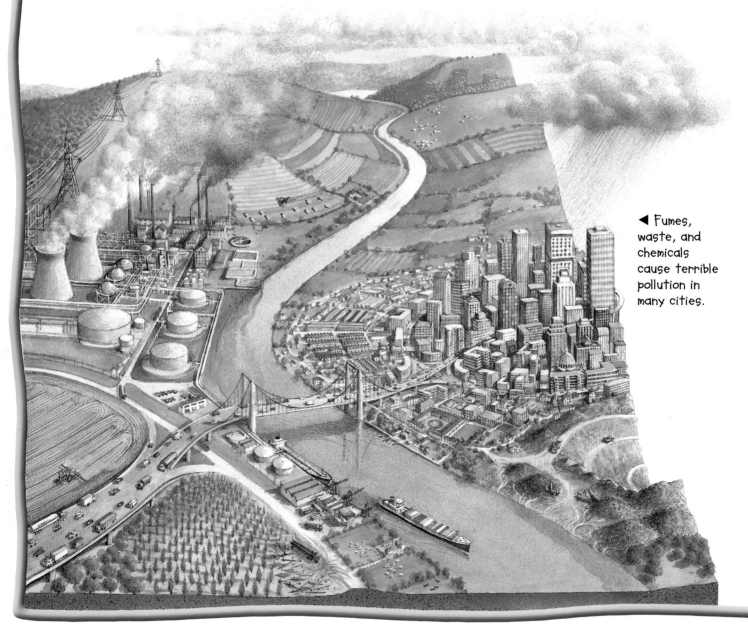

◄ Fumes, waste, and chemicals cause terrible pollution in many cities.

99

One very important area of science is recycling. Many materials and substances can be recycled—glass, paper, plastics, cans, scrap metals, and rags. Scientists are working to improve the process. Products should be designed so that when they no longer work, they are easy to recycle. The recycling process itself is also being made more effective.

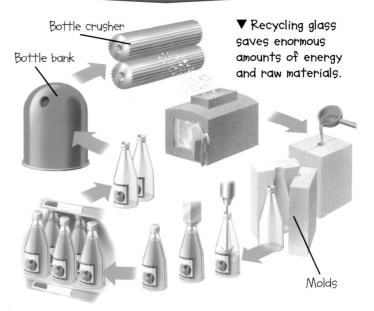

Bottle crusher

Bottle bank

▼ Recycling glass saves enormous amounts of energy and raw materials.

Molds

QUIZ

If you become a scientist, which science would you like to study? See if you can guess what these sciences are:
1. Meteorology 2. Biology
3. Astronomy 4. Ecology

1. Weather and climate
2. Animals, plants, and other living things 3. Stars, planets, and objects in space 4. The way nature works

▼ The energy in flowing water can be turned into electricity at a hydroelectric power station.

100

We use vast amounts of energy, especially to make electricity and as fuel in our cars. Much of this energy comes from crude oil (petroleum), natural gas, and coal. But these energy sources will not last forever. They also cause huge amounts of pollution. Scientists are working to develop cleaner forms of energy, which will produce less pollution and not run out. These include wind power from turbines, solar power from photocells, and hydroelectric and tidal power from dams.

Index

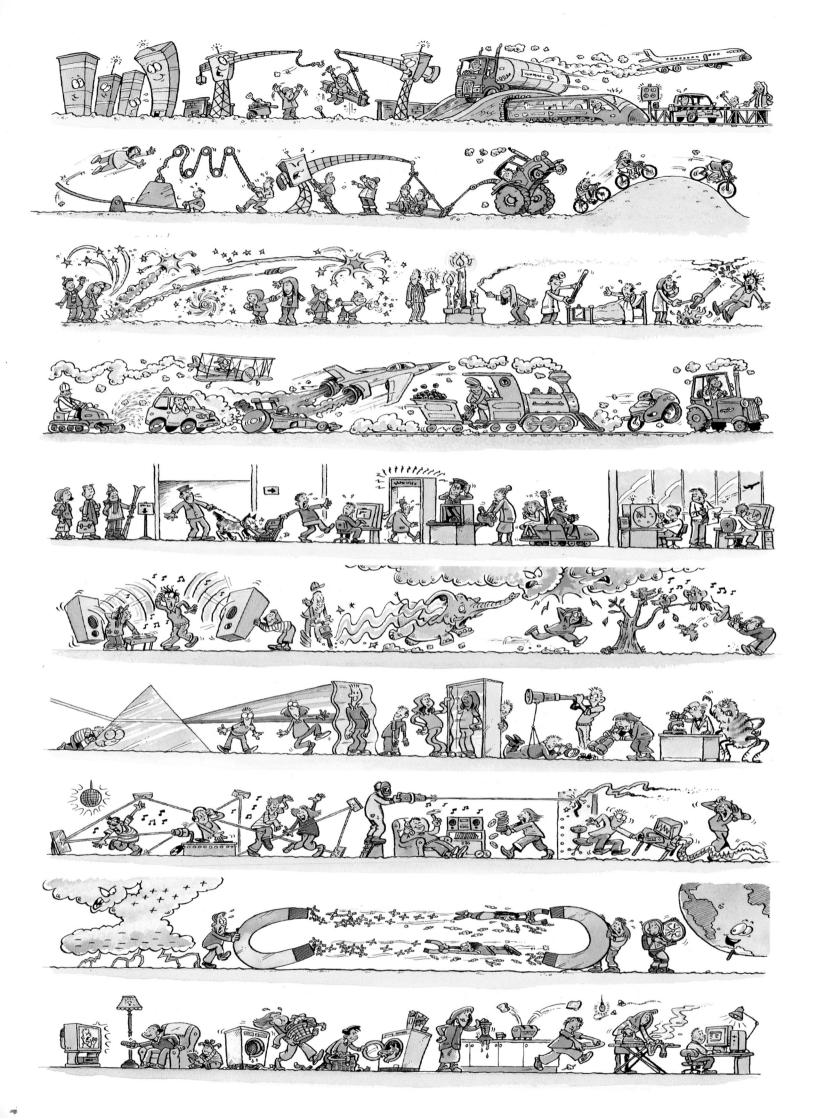